My STEM Workbook

1

Understanding
Science, Technology, Engineering and Mathematics through design-process activities

I ♥ STEM

This workbook belongs to

Years 1–2

Vinesh Chandra & Basil Slynko

My STEM Workbook 1
Understanding Science, Technology, Engineering and
Mathematics through design-process activities
Vinesh Chandra and Basil Slynko

Editor/Proofreader: Sandra Balonyi
Text designer: Michael Haddad
Cover designer: Michael Haddad
Illustrator: Michael Haddad

First published in Australia in 2023
Copyright ©2023 Vinesh Chandra and Basil Slynko
B. Slynko (Challenges 1, 5, 6, 8); V. Chandra (Challenges 2, 3, 4, 7)

National Library of Australia Cataloguing-in-Publication Data

Chandra, Vinesh and Slynko, Basil.
My STEM Workbook 1
Understanding Science, Technology, Engineering and Mathematics
through design-process activities

ISBN: 978-0-6484052-4-5
For primary school age.

Printed in Australia
1 2 3 4 5 6 7 8 9 29 28 27 26 25 24 23

Contents

Acknowledgements

The authors and publisher would like to credit or acknowledge the following sources for permission to use copyright material:

istock: p6 (gardening gloves/gumboots); p24 (roti/panini/ flatbreads); p28 (PPE); p36 (PPE); p40 (urban waste); p52 (PPE); p56 (bird house); P60 (PPE).

Shutterstock: p18 (bullock cart/hand cart); p25 (solar oven); p32 (globe); p56 (hanging bird shelter/natural log bird shelter).

Every attempt has been made to trace and acknowledge copyright holders. Where the attempt has been unsuccessful, the publisher welcomes information that would redress the situation.

The authors and publisher wish to thank the following individuals for the helpful advice they gave during the writing of this workbook: Julia Horn, Melissa Slynko, Charlene Cable-Lloy, Pauline Burrows-Booth, Alanah-Rei Castledine, Katie Whitehead, and Andrew Iddles.

A special "thank you!" to Sandra Balonyi and Michael Haddad for working with us on this project.

Responsibility for errors remains with the authors.

About the authors

Associate Professor Vinesh Chandra is a university lecturer and teacher with more than 40 years of experience. He has taught in Australia and several countries overseas. His teaching areas include STEM, science, mathematics and technologies. His co-authored book titled *STEM Education in the Primary School: A Teacher's Toolkit* received two awards (Educational Publishing Australia Awards, 2021). Associate Professor Chandra's team won Gold in the prestigious QS Reimagine Education Wharton Awards in 2022 for their project in STEM Education in Papua New Guinea.

Basil Slynko, aka Professor Baz, B Ed St., MA; Design and Technologies educator – primary, secondary and tertiary – in Australia and overseas; Project-based-learning advocate; Curriculum consultant; Industry Experience – Construction and Manufacturing; Author and co-author of 25 titles, including *Nelson Introducing Technology Fourth Edition, Nelson Technology Activity Manual Third Edition,* and *Design and Technology in Today's World: A First Look.*

Introduction

Humans have always relied on Science, Technology, Engineering and Mathematics, or STEM, to find solutions to challenges. Future generations will need to holistically draw upon this within and beyond their contexts. For example, STEM knowledge and skills are vital when addressing the United Nationals Sustainable Development Goals.

What is the United Nations?
Many years ago, nearly all countries in the world joined to form an organisation called the United Nations.

The United Nations aims to find ways to make our planet a better place for all.

*Member states and territories of the United Nations.

What are the United Nations Sustainable Development Goals?
The United Nations Sustainable Development Goals are about improving the lives of people and all other living things to make our planet a better place. They are also about what we can do to care for our Earth so that future generations can also have a happy and healthy life.

My STEM Workbook 1 is part of a trilogy of STEM workbooks for primary students Years 1–6. Each workbook has eight challenges. Each addresses one of the United Nations Sustainable Development Goals (SDGs). Students apply their knowledge and skills of Science, Technologies, Engineering and Mathematics to propose solutions to contextually appropriate real-world challenges. These activities align with a range of content descriptors mandated in the Australian Curriculum Science, Technologies, and Mathematics (Version 9). However, these activities can also be implemented in other contexts, guided by other curriculum documents.

Notes for the teacher

The essence of integrated STEM education is project-based learning (PBL). It is a fun way for students to learn and teach. The real-world challenge in each activity is highly likely to interest, engage and enthuse students. *My STEM Workbook 1* comprises eight hands-on design activities, where students apply their STEM knowledge and skills to propose solutions to real-world challenges informed by the United Nations Sustainable Development Goals (SDGs).

The STEM PBL framework[1] was used to design the activities. Students need to tackle each challenge through the following steps:

Ask → Imagine → Plan → Create → Improve

Through these steps students apply their design-thinking skills to propose solutions to real-world challenges. Workplace Health and Safety is an integral part of each activity. Students are expected to handle tools, equipment and materials with care. Teachers are also expected to reinforce the use of safety gear – that is, Personal Protective Equipment (PPE) – as needed.

The trilogy of STEM Workbooks has a website (https://mystemworkbook.com/) which presents ideas on how knowledge from the digital technologies curriculum can be embedded within each challenge. The website also has support materials, including a video commentary on each activity.

1. Forbes, A., Chandra, V., Pfeiffer, L., & Sheffield, R. (2021). *STEM education in the primary school: a teacher's toolkit*. Cambridge University Press.

Connections to the Australian Curriculum (Science, Design and Technologies, and Mathematics)

Subject	Content Descriptions	Activity							
		1	2	3	4	5	6	7	8
SCIENCE	identify the basic needs of plants and animals, including air, water, food or shelter, and describe how the places they live meet those needs (AC9S1U01)	×	×	×	×			×	×
	describe daily and seasonal changes in the environment and explore how these changes affect everyday life (AC9S1U02)	×			×				
	describe pushes and pulls in terms of strength and direction and predict the effect of these forces on objects' motion and shape (AC9S1U03)			×					
	describe how people use science in their daily lives, including using patterns to make scientific predictions (AC9S1H01)	×	×	×	×	×	×	×	×
	pose questions to explore observed simple patterns and relationships and make predictions based on experiences (AC9S1I01)			×					×
	suggest and follow safe procedures to investigate questions and test predictions (AC9S1I02)	×	×	×	×	×	×	×	×
	sort and order data and information and represent patterns, including with provided tables and visual or physical models (C9S1I04)	×	×	×	×				
	compare observations with predictions and others' observations, consider if investigations are fair and identify further questions with guidance (AC9S1I05)					×	×	×	×
	write and create texts to communicate observations, findings and ideas, using everyday and scientific vocabulary (AC9S1I06)	×	×	×	×				
	explore different actions to make sounds and how to make a variety of sounds, and recognise that sound energy causes objects to vibrate (AC9S2U02)						×		
	recognise that materials can be changed physically without changing their material composition and explore the effect of different actions on materials including bending, twisting, stretching and breaking into smaller pieces (AC9S2U03)					×		×	×
	describe how people use science in their daily lives, including using patterns to make scientific predictions (AC9S2H01)	×	×	×	×	×	×	×	×
	pose questions to explore observed simple patterns and relationships and make predictions based on experiences (AC9S2I01)	×	×	×	×	×	×	×	×
	suggest and follow safe procedures to investigate questions and test predictions (AC9S2I02)	×	×	×	×	×	×	×	×
	sort and order data and information and represent patterns, including with provided tables and visual or physical models (AC9S2I04)	×	×	×	×	×	×		×
	compare observations with predictions and others' observations, consider if investigations are fair and identify further questions with guidance (AC9S2I05)	×	×	×	×	×	×	×	×
	write and create texts to communicate observations, findings and ideas, using everyday and scientific vocabulary (AC9S2I06)	×	×	×	×	×	×	×	×

Subject	Content Descriptions	Activity							
		1	2	3	4	5	6	7	8
DESIGN AND TECHNOLOGIES	identify how familiar products, services and environments are designed and produced by people to meet personal or local community needs and sustainability (AC9TDE2K01)				×	×	×	×	×
	explore how technologies including materials affect movement in products (AC9TDE2K02)			×					
	explore how plants and animals are grown for food, clothing and shelter (AC9TDE2K03)	×							
	explore how food can be selected and prepared for healthy eating (AC9TDE2K04)		×		×				
	generate and communicate design ideas through describing, drawing or modelling, including using digital tools (AC9TDE2P01)	×	×	×	×	×	×	×	×
	use materials, components, tools, equipment and techniques to safely make designed solutions (AC9TDE2P02)	×	×	×	×	×	×	×	×
	evaluate the success of design ideas and solutions based on personal preferences and including sustainability (AC9TDE2P03)	×	×	×	×	×	×	×	×
	sequence steps for making designed solutions cooperatively (AC9TDE2P04)	×	×	×	×	×	×	×	×
	recognise, represent and order numbers to at least 120 using physical and, numerals, number lines and charts (AC9M1N01)	×	×	×	×	×	×	×	×
	use mathematical modelling to solve practical problems involving situations including simple money transactions; represent the situations with diagrams, physical and virtual materials, and use calculation strategies to solve the problem (AC9M1N05)	×	×	×	×	×	×	×	×
MATHEMATICS	use mathematical modelling to solve practical problems involving equal sharing and grouping; represent the situations with diagrams, physical and virtual materials, and use calculation strategies to solve the problem (AC9M1N06)	×	×	×	×	×	×	×	×
	compare directly and indirectly and order objects and events using attributes of length, mass, capacity and duration, communicating reasoning (AC9M1M01)	×	×	×	×	×	×	×	×
	measure the length of shapes and objects using informal units, recognising that units need to be uniform and used end-to-end (AC9M1M02)	×	×	×	×	×	×	×	×
	make, compare and classify familiar shapes; recognise familiar shapes and objects in the environment, identifying the similarities and differences between them (AC9M1SP01)	×	×	×	×	×	×	×	×
	give and follow directions to move people and objects to different locations within a space (AC9M1SP02)							×	
	recognise, represent and order numbers to at least 1000 using physical and virtual materials, numerals and number lines (AC9M2N01)	×	×	×	×	×	×	×	×
	recognise and describe one-half as one of 2 equal parts of a whole and connect halves, quarters and eighths through repeated halving (AC9M2N03)		×						
	use mathematical modelling to solve practical problems involving additive and multiplicative situations, including money transactions; represent situations and choose calculation strategies; interpret and communicate solutions in terms of the situation (AC9M2N06)	×	×	×	×	×	×	×	×
	measure and compare objects based on length, capacity and mass using appropriate uniform informal units and smaller units for accuracy when necessary (AC9M2M01)	×	×	×	×	×	×	×	×
	identify common uses and represent halves, quarters and eighths in relation to shapes, objects and events (AC9M2M02)		×						
	recognise, compare and classify shapes, referencing the number of sides and using spatial terms such as "opposite", "parallel", "curved" and "straight" (AC9M2SP01)	×	×	×	×	×	×	×	×
	locate positions in two-dimensional representations of a familiar space; move positions by following directions and pathways (AC9M2SP02)	×	×	×	×	×	×	×	×
	acquire data for categorical variables through surveys, observation, experiment and using digital tools; sort data into relevant categories and display data using lists and tables (AC9M2ST01)	×	×	×	×	×	×	×	×

Challenge Activities

Notes for the student

In this workbook, you will engage in eight activities. In each activity, you will use science, technology, engineering and mathematics, or STEM, to develop solutions to real-world challenges.

What will I do in the Challenge activities?

In each challenge activity you will follow these steps:

Step 1. Ask
What is the challenge?

Step 2. Imagine
How can I tackle the challenge?

Step 3. Plan
How can I plan my idea?

Step 4. Create
What will my idea look like?

Step 5. Improve
How can I make my idea better?

Then, at the end, you will **reflect** on the activity and its outcomes to see what you have learnt.

Now, let's have a look at the eight Challenge activities. Each activity is based on a different United Nations Sustainable Development Goal...

Challenge 1
– plan a vegetable garden
Goal: Zero Hunger

Challenge 2
– plan a recipe for a healthy sandwich
Goal: Good Health and Well-being

Challenge 3
– plan a water transporter
Goal: Clean Water and Sanitation

Challenge 4
– plan a portable solar oven
Goal: Affordable and Clean Energy

Challenge 5
– plan a waste-storage area
Goal: Sustainable Cities and Communities

Challenge 6
– plan a musical instrument made from waste
Goal: Climate Action

Challenge 7
– plan a board game about waterways
Goal: Life Below Water

Challenge 8
– plan a shelter for native birds
Goal: Life on Land

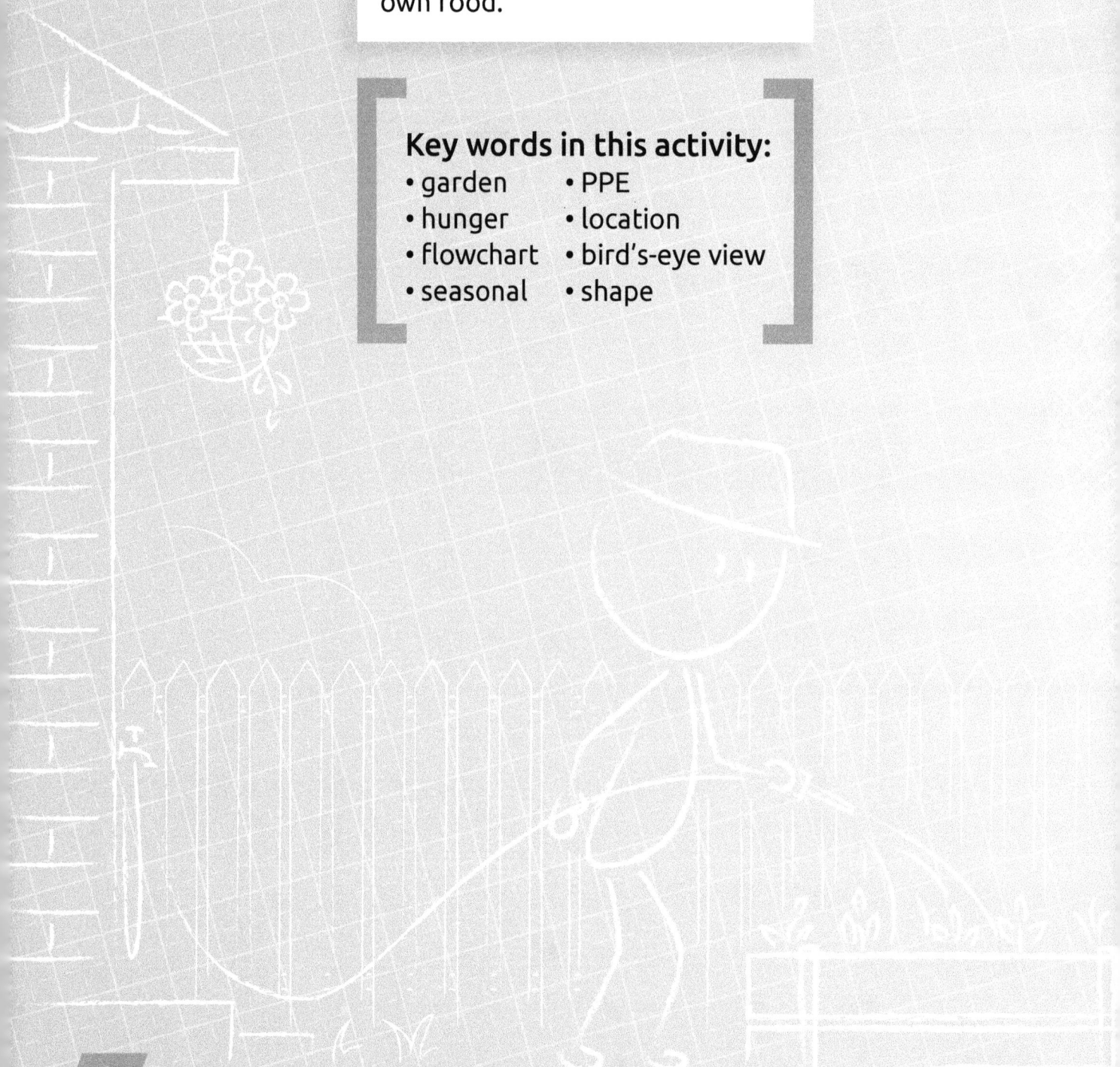

Zero Hunger

1

There are many people in the world who do not have enough to eat. Zero **Hunger** is one of the goals of the United Nations. The Zero Hunger Goal is about finding ways to make sure that everyone has enough to eat. One of the ways to do this is by growing your own food.

Key words in this activity:
- garden
- hunger
- flowchart
- seasonal
- PPE
- location
- bird's-eye view
- shape

Challenge 1 Plan a vegetable garden

Challenge
– plan a vegetable garden

Your school needs a vegetable garden. Your challenge is to come up with a plan for a vegetable garden. The garden should be about the size of your bed.

1. Ask...

Here are three vegetable **garden** ideas: a rectangular tray, a window box and a raised garden bed.

Why should we grow our own vegetables?

A trellis

A rectangular trough

Q1. Compare the three vegetable garden ideas. Which garden idea is the best? Why? *Write your answers here.*

The best garden idea is

This garden idea is the best because

A window box

A raised garden bed

2. Imagine...

Did you know that some vegetables need to be grown in certain ways?

Some may need:
- a trellis to climb (beans)
- more space (pumpkin)
- deeper soil (potatoes).

Challenge 1 Plan a vegetable garden

Q2a. Some things to consider when planning a vegetable garden are **location**, soil, access to water and size. Why are these things important?
Write your answers here.

The location of the garden is important because	The type of soil is important because
Access to water is important because	This size of the garden is important because

Q2b. Some vegetables are **seasonal**. What does this mean?

Q2c. What vegetables do you want in the garden? Share your ideas with your classmates. What do they think? *Write their comments here.*

Q2d. *List some vegetables that can be grown in the vegetable garden.* What do they need to grow?

Vegetable name	What it needs to grow
1.	
2.	
3.	
4.	

Q2e. *List three materials that could be used to build a garden.*

Material 1:
_____

Material 2:
_____

Material 3:
_____

3. Plan...

Q3a. Use the grid to *draw a rough* **bird's-eye view** *of your initial garden idea.*

Q3b. Share your garden idea with your classmates. What do they think? *Write their comments here.*

Q3c. Use some of the classmates' ideas to improve your garden. *Now, draw a final bird's-eye view of your garden.*

Remember to:

- label the materials and the vegetables you want to grow
- show distance between plants using paddle-pop sticks as a measure.

The **shape** *of my garden is* _____.

Q3d. In the table, list the plants and quantities (No.) you will need for your vegetable garden.

Plant	No.	Plant	No.

4

Procedure

Q3e. How will you make your garden? *Follow the steps from 1 to 9 and fill in the blanks in the* **flowchart** *using the words from the Word Bank. Use arrows to show the connection between the steps.*

Did you know...
A flowchart is a diagram showing the steps needed to complete a task. Arrows are used to connect the steps.

Word Bank

build   layout   materials   mulch
plant   planting   site   soil   water

Select a ☐ 1

Mark out the ☐ 2

Get some ☐ 3

Get the ☐ 4

☐ 5 the frame.

Mark out the ☐ 6

☐ 7 the vegetables.

Add ☐ 8 to the garden.

☐ 9 the garden.

Challenge 1 Plan a vegetable garden

Tools and Equipment

Safety is important to prevent accidents and injuries. Using tools and equipment incorrectly can be dangerous. You should always use safety gear to protect yourself when using tools and equipment. The other term for safety gear is "Personal Protective Equipment", or **"PPE"**.

Gardening tools are used to work the vegetable garden. Some gardening tools are rakes, shovels, garden forks, hoes, saws and secateurs.

Some examples of PPE: safety glasses, gardening gloves and gumboots.

Rake

Shovel

Saw

Secateurs

Hoe

Garden fork

Q3f. *Select three gardening tools and write them in the "Gardening tool" column of the table below.*

Gardening tool	How I could get injured	PPE I will need
1.		
2.		
3.		

Q3g. *Now complete the other two columns in the table.*
• Each tool can injure you. Write down how they can injure you.
• What PPE will you need?

Challenge 1 Plan a vegetable garden

4. Create... [Optional]

Making a vegetable garden

You need to:
- gather the materials for your vegetable garden
- gather the tools and equipment you need
- follow your flowchart steps to make the vegetable garden
- use PPE and follow the safety rules.

5. Improve...

Present your final garden idea to your class. Explain your choice of plants.

Ask your classmates these questions:
- What are some good points about my vegetable-garden idea?
- What are some weak points about my vegetable-garden idea?

Q5a. *Write your classmates' answers in the table.*

Good points	Weak points

Q5b. Consider your classmates' answers. *List two things you could do to improve your garden idea.*

Now it's time to Reflect...

What have you learnt in this activity? *List three things.*
Hint: Look back at the key words.

1.

2.

3.

What would you like to know more about?

2

Good Health and Well-being

To enjoy our lives, we all have to be **healthy** and well. Good Health and Well-being is also a goal of the United Nations. Our health depends on the food we eat.

Key words in this activity:
- healthy
- fraction
- food safety
- procedure

Challenge
- plan a recipe for a healthy sandwich

To mark Children's Week you are going on a picnic. Your mum would like you to make two healthy sandwiches using four slices of bread that can be packed for lunch. Your challenge is to come up with an idea for making healthy sandwiches.

1. Ask...

Q1a. Name some healthy foods.

Q1b. Why do we need to eat healthy foods?

Q1c. What is a healthy sandwich?

Q1d. *Draw lines on the slice of bread to show how you would cut it into equal pieces so it is easy to pack.* Label each piece of bread as a **fraction**.

Each piece looks like a

_____

2. Imagine...

Q2a. Look at the photo above. *Tick the ingredients that you will use to make the sandwiches. Then write down some other ingredients you could use.*

Other ingredients I could use are:

Q2b. *Jot down some ideas for making a healthy sandwich and share them with your class.*

Challenge 2 Plan a recipe for a healthy sandwich

3. Plan...

Q3a. What ingredients will you need to make healthy sandwiches for your family? *In the table, list the ingredients and the quantities you will need. You also need to say why you will need each ingredient.*

Ingredients	Quantity	Why do I need this ingredient?

Procedure

3b. How will you make the sandwiches? *Write down the steps you will follow.* You can have as many steps as you like.

Remember: One step has to say how you would cut the sandwich into equal pieces.

Tools and Equipment

Safety is important to protect yourself from cuts and injuries. Using kitchen utensils and equipment can be dangerous. You should always think about **food safety** when handling food.

Q3c. You will need some kitchen utensils and equipment to make a sandwich. Look at the kitchen utensils and equipment in this photo. Which ones will you need? *Fill in the blanks below with the name of three utensils and what you will use them for.*

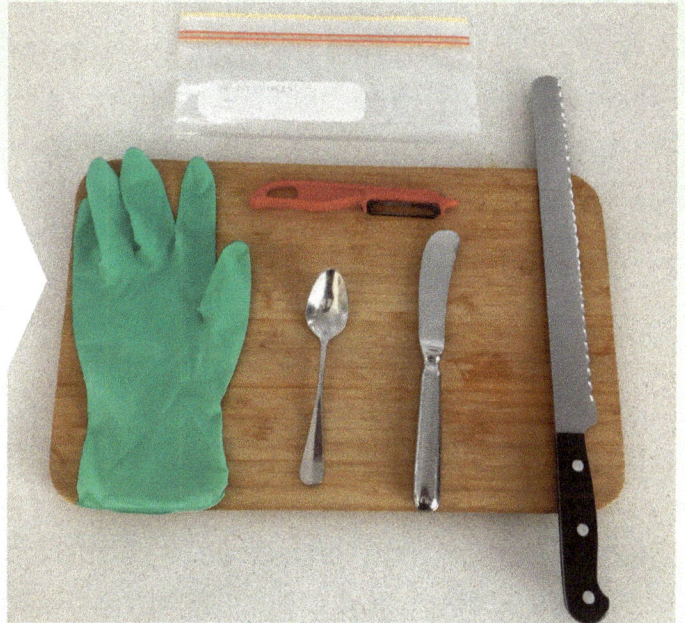

Get serious bout food safety!

Food safety is in your hands!
1. Be clean, be healthy.
2. Keep food hot or cold.
3. Wash, rinse and sanitise.

1. *I will need a* _____

   to _____ .

2. *I will need a* _____

   to _____ .

3. *I will need a* _____

   to _____ .

Q3d. You need to follow some food-safety rules. Which of these food-safety rules are important when you make a sandwich?

- [ ] Wash your hands with soap.
- [ ] Dry your hands.
- [ ] Put on some sunscreen.
- [ ] Put on a pair of gloves.
- [ ] Be careful when using a knife to cut bread.

4. Create... [Optional]

Making sandwiches

You need to:
- gather the ingredients for your sandwiches
- gather your kitchen utensils and equipment
- follow your **procedure** for making sandwiches
  (don't forget to cut the sandwiches into equal pieces)
- get feedback from your teacher.

5. Improve...

Present your sandwich idea to your classmates. Explain why your sandwich is healthy.

What did your class think of your sandwich idea? Some questions to ask them are:
- What are the good points about my sandwiches?
- What are the weak points about my sandwiches?

Q5a. *Write your classmates' answers in the table.*

Good points	Weak points

Q5b. Consider your classmates' comments. How could you improve your sandwiches next time?

Now it's time to Reflect...

What have you learnt in this activity? *List three things.*
Hint: Look back at the key words.

1.

2.

3.

What would you like to know more about?

14

3

Clean Water and Sanitation

We all need water that is clean. In some parts of the world **clean water** is not available. One of the goals of the United Nations is to find ways for everyone to easily access clean water and sanitation.

Key words in this activity:
- clean water
- water transporter
- force
- PPE

Challenge
- plan a water transporter

Water in many parts of the world needs to be transported to people. Your challenge is to plan a model of a water transporter. Your model should be built from recyclable materials and should be no more than one ruler length long.

1. Ask...

Q1a. Why does water need to be transported to people in many parts of the world?

Q1b. What is clean water?

We use water in different ways.

For example, you might fill your water bottle and drink from it when you are in the classroom.

Q1c. Why do we need clean water?

2. Imagine...

This is Epi. Epi believes he drinks two bottles of water each day. The table below shows how much water Epi uses for different activities each day.

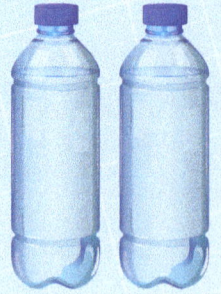

Activity	Water used
1. Washing hands and brushing teeth	5 bottles
2. Flushing the toilet	10 bottles
3. Washing clothes	15 bottles
4. Watering the garden	20 bottles
5. Taking a shower	30 bottles

Q2a. How much water does Epi use each day? *Write your answer here.* ➜   _____ bottles

Q2b. *Complete this tally chart to show the amount of water used by Epi.* The first line of the chart has been completed as an example.

Activity	Water used	Tally
1. Washing hands and brushing teeth	5 bottles	ⅲ̶
2.	bottles	
3.	bottles	
4.	bottles	
5.	bottles	

A **force** *is needed to move a thing. Pushing and pulling are two common examples. Arrows are used to show the direction of a force.*

Pushing
→

Pulling
→

In some parts of the world water needs to be transported. These photos show two examples.

Q2c. What force is required in each example and why?

Bullock cart: *force needed is* _____.

Why?

Hand cart: *force needed is* _____.

Why?

Q2d. What recyclable materials can you use to build your **water-transporter** model? *Jot down some ideas.*

3. Plan...

Q3a. Use the grid to *draw a side view of your initial water-transporter model.*

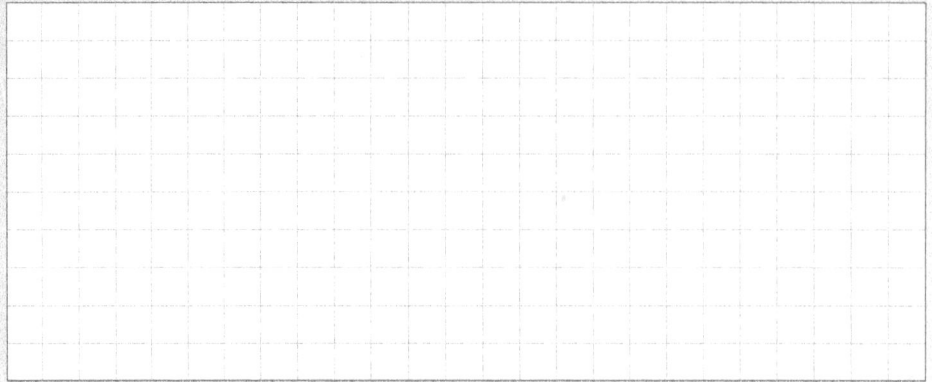

Q3b. Share your water-transporter idea with your classmates. What do they think? *Write their comments here.*

Q3c. Consider the comments. *Now draw your final idea.*
In your drawing, you need to:
- label the parts of the model
- show the force needed to move your model
- add any other information and details.

Q3d. *In the table, list the recyclable materials and the quantities you will need to build your model. You also need to say why you will need each material.*

Material	Quantity	Why do I need this material?

My STEM Workbook 1 – Understanding Science, Technology, Engineering and Mathematics through design-process activities   Vinesh Chandra and Basil Slynko   ISBN: 978-0-6484052-4-5

Procedure

Q3e. *Write down the steps you will follow to make your model.*
You can have as many steps as you like.

Tools and Equipment

Safety is important to prevent accidents and injuries. Using tools and equipment can be dangerous. You should use safety gear when making your model. The other term for safety gear is "Personal Protective Equipment", or "**PPE**".

Q3f. *List the tools and equipment you will need to make your model, and what use they will have.*

Tool / equipment	Use

Q3g. *List three safety rules you will need to follow when making your model.*

Challenge 3 Plan a water transporter

4. Create... [Optional]
Making a water transporter

You need to:
- gather all the materials, tools and equipment you will need
- follow the procedure to make the model
- get feedback from your teacher.

5. Improve...

Present your water-transporter idea to your class. Explain what force will be needed to move it.

Ask your classmates these questions:
- What are some good points about my water-transporter idea?
- What are some weak points about my water-transporter idea?

Q5a. *Write your classmates' answers in the table.*

Good points	Weak points

Q5b. Consider your classmates' answers. What could you do to improve your water-transporter idea next time?

Now it's time to Reflect...

What have you learnt in this activity? *List three things.*
Hint: Look back at the key words.

1.

2.

3.

What would you like to know more about?

4

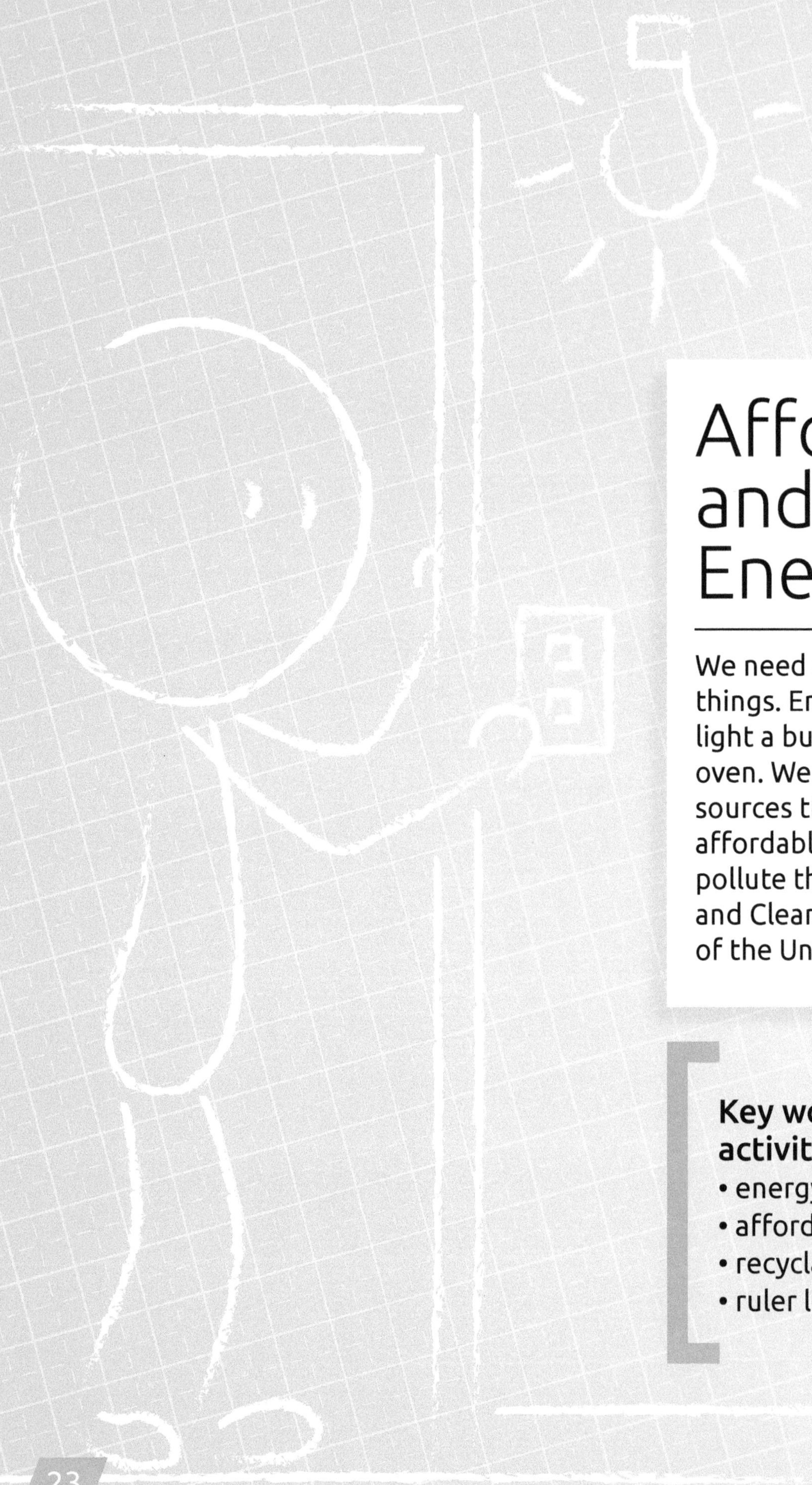

Affordable and Clean Energy

We need **energy** to do things. Energy to move, light a bulb and heat the oven. We need energy sources that are clean and affordable. They should not pollute the air. Affordable and Clean Energy is a goal of the United Nations.

Key words in this activity:

- energy
- affordable
- recyclable materials
- ruler length
- clean
- PPE

Challenge
- plan a portable solar oven

Your teacher has set an energy challenge. Using the sun as an energy source, you have to come up with an idea for making a portable solar oven using recycled materials. Your oven should be no larger than the size of a takeaway pizza box. Your oven should be able to warm food such as rotis, flatbread and paninis.

1. Ask...

Q1a. What is solar energy?

Roti, paninis and flatbreads

Q1b. What is a solar oven?

The sun is one alternative energy source. Some other alternative energy sources are...

Tidal Power

Hydroelectric Power

Q1c. *Complete the following sentences:*

Solar energy is **clean** because

Wind Power

Solar energy is **affordable** because

Geothermal Power

24

2. Imagine...

Q2a. Look at this photo of a solar oven. How does it work?

Q2b. Are these statements **true** or **false**? *Circle your answer.*

i)   The oven will work best in the evening.   True | False

ii)   The oven will work best on a sunny day.   True | False

iii) The oven will work best when it is raining.   True | False

iv)   The oven will work best in the shade.   True | False

v)   The oven will work best inside the house.   True | False

Q2c. What **recyclable materials** can you use to build a portable solar oven?

Challenge 4 Plan a portable solar oven

3. Plan...

Q3a. Use the grid to *draw a rough bird's-eye view of your portable solar-oven idea.*

Q3b. Share your portable solar-oven idea with your classmates. What do they think of your idea? *Write their comments below.*

Consider the feedback and review your idea...

Q3c. *Now draw your final idea on the grid.* Remember to:
- label the parts of the solar oven (for example: base)
- label the materials
- add any other information and details.

*State the length and width in **ruler lengths**:*

The length of the oven = _____ ruler length(s).

The width of the oven = _____ ruler length(s).

26

Q3d. *In the table, list the materials and the quantities you will need to make a portable solar oven. You also need to say why you will need each material.*

Material	Quantity	Why do I need this material?

Procedure

Q3e. How could you make a solar oven? *Write the steps you will take.* You can have as many steps as you like.

Find-a-word

Can you find these words in the find-a-word?

AFFORDABLE   PORTABLE
CLEAN   RECYCLABLE
ENERGY   SAFE
FUEL   SOLAR
HEAT   SUN
MATERIALS   SUSTAINABLE

```
B O F G J X H A U F M N S U
W A Z R N Q E S T P U S O C
T F R O D S A W V N M E R I
P F O S U S T A I N A B L E
S O L A R X V S Y B T I J N
W R E C Y C L A B L E R G E
R D P C L V A F S X R K L R
K A C H Y E B E M S I B O G
D B E L R E A Y Z O A C P Y
E L D Q U D F N G T L J T Y
Y E R F G K A I R O S U N D
W S T I Q B V O W K M N A I
C V S Z G X P D Q I S O F M
```

Tools and Equipment

Safety is important to prevent accidents and injuries. Using tools and equipment can be dangerous. You should use safety gear when using tools and equipment. The other term for safety gear is "Personal Protective Equipment", or "**PPE**".

Some examples of PPE: gloves, an apron and face masks.

Q3f. *Draw the tools and equipment you will need to make your portable solar oven.*

Q3g. *List three safety rules you will need to follow when making your portable solar oven.*

SAFETY RULE 1	SAFETY RULE 2	SAFETY RULE 3

4. Create... [Optional]

Making a solar oven

You need to:
- gather all the materials, tools and equipment you will need
- follow your procedure to make your solar oven
- get feedback from your teacher.

5. Improve...

Present your solar-oven idea to your class. Explain how it works and what makes it affordable and clean.

Questions to ask are:
- What are the good points of my solar-oven idea?
- What are the weak points of my solar-oven idea?

Q5a. *Write your classmates' answers in the table.*

Good points	Weak points

Q5b. Consider your classmates' comments. How could you improve your solar oven next time?

Now it's time to Reflect...

What have you learnt in this activity? *List three things.*
Hint: Look back at the key words.

1.

2.

3.

What would you like to know more about?

5

Sustainable Cities and Communities

One of the goals of the United Nations is about taking good care of our cities and communities. We can all have a better life when we look after the cities and communities where we live. One of the ways to do this is through the sustainable use of resources. Sustainable use means using resources in a way that meets the needs of people now as well as future generations. In this way, we also care for our Earth.

RECYCLE → REDUCE → REUSE → RECYCLE → REDUCE → REUSE →

NO WASTE

Key words in this activity:
- sustainability
- sustainable
- no waste
- recycle
- PPE
- reduce
- reuse
- hazard

Challenge 5 Plan a waste-storage area

Challenge
- plan a waste-storage area

The local council has asked primary school children to think about **sustainable cities and communities.** The goal is to encourage sustainable habits in the community. Your class has been asked to develop a plan of action to encourage sustainable actions of "no waste" and "recycle, reduce and reuse" activities at home. Your challenge is to plan a waste-storage area for your home or school.

1. Ask...

Q1a. What does **sustainability** mean to you?

Q1b. What actions do you take to manage waste? *List three actions.*

Action 1:

Action 2:

Action 3:

Q1c. Are the actions you listed sustainable? *Circle your answer then list why.*

Action 1: Yes | No | Maybe   Why?

Action 2: Yes | No | Maybe   Why?

Action 3: Yes | No | Maybe   Why?

2. Imagine...

Have you thought about sustainable actions? What are some of the things we can all do to make our planet a better place to live in?

This is Sarah. She wrote the following:

- I can do something about household waste.
- I can do something about water usage.
- I can also do something about recycling.
- I can also do something about how I use plastics.
- I can also do something about green waste.

Q2a. What can Sarah do? *Write some ideas for each heading.*

Household waste:

Water usage:

Recycling:

Plastics:

Green waste:

Sustainable habits are easier to keep up if there is a process of getting rid of waste. One way is to have an area set aside for storage of waste such as household, recycling (cans and bottles), plastics and green.

Q2b. What materials could you use to hold the waste? *Complete the table.* →

Type of waste	Suitable materials to hold waste
Household	
Recycling (cans and bottles)	
Plastics	
Green	

Challenge 5 Plan a waste-storage area

3. Plan...

Q3a. Use the grid to *draw a bird's-eye-view of your initial idea for short-term storage.*

Q3b. Share your waste-storage idea with a few classmates. What do your classmates think of your idea? *Write their comments here.*

Consider the feedback and review your idea...

Q3c. Use some of the ideas suggested by your classmates to improve your waste-storage idea. *Then draw your final idea on the grid.* Remember to:
- label the materials, equipment and position of waste to be stored
- add any other information and details.

Procedure

Q3d. How could you set up a waste-storage area? *List the steps.*
You can have as many steps as you like.

The right path

Sarah wants to take her rubbish to the waste-storage area. *Can you help her find the right path?*

Challenge 5 Plan a waste-storage area

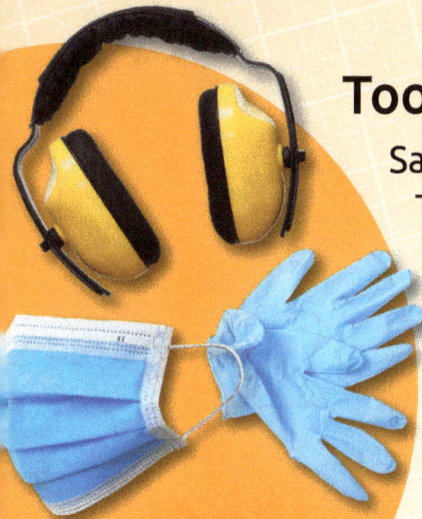

Tools and Equipment

Safety is important to prevent accidents and injuries. There is a safety risk when recycling waste. There are also **hazards** to be aware of when handling and storing waste. **"PPE"** (Personal Protective Equipment) is the general term used to describe safety gear. You should use PPE to protect yourself when handling and storing waste.

Some examples of PPE: ear muffs, face mask and disposable gloves.

Q3e. A range of waste is listed in the first column of the table. *Complete the other two columns.*

• What are the hazards of handling each type of waste?

• What PPE will you need?

Waste	The hazard is	PPE I will need
Household waste		
Recycling • Cans		
• Plastics		
• Paper and cardboard		
• Glass		
Green waste		

36

4. Create... [Optional]

Making a model of your waste-storage area

You need to:

- gather all the materials: paper, cardboard and paddle-pop/ice-cream sticks
- gather the tools: scissors, stapler or bulldog clips
- follow the procedure to make the storage area
- use PPE and follow the safety rules
- get feedback from your teacher.

5. Improve...

Present your waste-storage idea to your class.

Questions to ask are:

- What are some good points about my short-term waste storage idea?
- What are some weak points about my short-term waste storage idea?

Q5a. *Write your classmates' answers in the table.*

Good points	Weak points

Q5b. Consider your classmates' answers. What could you do to improve your short-term waste storage idea next time?

Now it's time to Reflect...

What have you learnt in this activity? *List three things.*
Hint: Look back at the key words.

1.

2.

3.

What would you like to know more about?

Climate Action

6

Action on **climate change** is one of the goals of the United Nations. What we do on Earth affects the climate. A huge amount of waste material is being dumped every day around the world. This causes **pollution**, which can affect the world's climate. Many materials can be reused and recycled. New items can be made so that our Earth is not polluted with waste materials.

Key words in this activity:
- climate change • PPE
- recycling • shapes
- pollution • percussion

39

Challenge
- plan a musical instrument made from waste

Planet Ark is an environmental organisation that holds a National Recycling Week in November each year. The local **recycling** centre is asking primary-school students under the age of eight to plan a musical instrument that you can strike or shake using a range of recyclable materials. The musical instrument should be about the size of your lunchbox. Your musical instrument may be displayed at the recycling centre.

Urban waste

1. Ask...

Q1a. *Fill in the blanks* using words from the Word Bank.

Dumping wastes causes _____.

We can reduce pollution by _____ *some of the waste.*

In this way we can help the world's _____.

Word Bank

climate
pollution
recycling

Musical instruments can fall into four groups. They are...

Q1b. *Add another musical instrument to each musical group.* Use the Internet or discuss with your classmates.

1 brass
- tuba or

2 percussion
- cymbals or

3 strings
- violin or

4 woodwind
- flute or

Sounds of musical instruments are produced by actions such as blowing, plucking, shaking or striking.

Q1c. Ed has learnt to play lots of musical instruments. Can you name the instruments and the actions that Ed would use to produce the sounds?
Fill in the correct number of the instrument and the action from the Word Bank.

Instrument number	Action

Word Bank

Instrument:
1. drum 2. guitar 3. harp
4. maracas 5. saxophone
6. tambourine 7. triangle
8. trumpet 9. violin
10. xylophone

Action:
blow pluck shake strike

Challenge 6 Plan a musical instrument made from waste

2. Imagine...

Producing sounds

Sound changes:

- with each different material, such as timber, metal, glass or plastics
- according to the size of the material – length, width and thickness
- according to the **shape** of the material, such as flat (2D) or folded (3D)
- if the item is solid or hollow
- if hollow items are filled with different amounts of materials, such as liquid, rice or pebbles
- when an action varies on a **percussion** instrument, for example a strike or a tap.

> **Did you know**
> *that you can use recyclable materials to make musical instruments?*

Q2a. Select different recycled materials and produce sounds using different actions. Use the information above to explore sounds. What did you find?

Q2b. What recycled materials could you use to make your musical instrument? *List them here.*

3. Plan...

Q3a. Use the grid to *draw a view of your initial idea for a instrument that you can shake or strike.*

Q3b. Share your initial musical-instrument idea with your classmates. What do they think of your initial idea? Write their comments below.

Q3c. Consider the comments. Review your initial idea. *Now draw your final idea on the grid.* Remember to:
- list the parts of your musical instrument
- label the recycled material(s) for each part
- record any other details.

The musical instrument group is _____.

Q3d. Complete the material list below.

Part of musical instrument	Quantity	Recyclable material is

Procedure

Q3e. How will you make your musical instrument? *Write down the steps you will follow.* You can have as many steps as you like.

Tools and Equipment

There is a safety risk when using tools and equipment. "**PPE**" (Personal Protective Equipment) is the general term used to describe safety gear. You should wear PPE to protect you when using tools and equipment.

Q3f. What PPE will you need when handling the recyclable materials listed in this table?

Material	PPE I will need
Glass	
Broken concrete	
Steel and tin cans	

Safety is important to prevent accidents and injuries.

4. Create... [Optional]
Making a musical instrument

You need to:
- gather the recyclable materials for your musical instrument
- gather the tools and equipment
- follow your procedure to make the musical instrument
- use PPE and follow the safety rules
- get feedback from your teacher.

5. Improve...

The next step is to present your musical-instrument idea to your class. Explain how it will work.

Questions to ask are:
- What are the good points of the musical instrument made from recyclable materials?
- What are the weak points of the musical instrument made from recyclable materials?

Q5a. *Write your classmates' answers in the table.*

Good points	Weak points

Q5b. Consider your classmates' comments. How could you improve your musical instrument next time?

Now it's time to Reflect...

What have you learnt in this activity? *List three things.*
Hint: Look back at the key words.

1.

2.

3.

What would you like to know more about?

46

7

Life Below Water

Every day people throw out **rubbish** that enters our ponds, lakes, streams, rivers and oceans. Rubbish in our waterways can be very dangerous for animals living in the water. One of the goals of the United Nations is for everyone in the world to work at finding ways to help animals below water to survive.

Key words in this activity:
- rubbish
- figurine
- squares
- visual appeal
- PPE

Challenge 7 Plan a board game about waterways

Challenge
- plan a board game about waterways

Your teacher wants you to plan a board game that will help people understand more about what rubbish can do to the animals who live in the water. The board game should not be larger than the size of your desk.

1. Ask...

Q1a. What are the needs of animals that live below water?

Q1b. How does rubbish get into waterways?

2. Imagine...

Q2a. The table below has some examples of rubbish that is found in our waterways. How can this rubbish affect the lives of animals below water?

Rubbish	How it can affect the lives of animals
Plastic bottles	
Plastic bags	
Fishing nets	

Q2b. What other rubbish is found in our waterways and oceans?

Q2c. What is your favourite animal that lives below water?

Q2d. What can rubbish do to your favourite animal?

Planning a board game

Board games are usually made up of **squares**. It is common to have two colours. One colour is used for the background. The other colour shows the squares that are "no go". "No go" can mean you miss a turn or move back a number of squares. The use of colours and the way the squares are arranged adds to the look of the game. This look will change if the "no go" squares are moved to form a different layout. The overall look is known as **visual appeal**.
Note: How many squares you have on the board game is up to you.

This is how Margaret made her board game...

She used blue for clean water. Animals move through clean water easily. She used red for water that was polluted with rubbish. Animals find it difficult to move through water with rubbish in it.

Margaret wrote the following rules:
1. Each player needs a **figurine** and a dice to play.
2. Roll the dice and move the figurine the number of squares shown on the dice.
3. Your figurine is safe when it lands on a blue square.
4. The figurine is in danger when it lands on a red square. When this happens, the figurine has to move back one square.
5. The figurine that gets to 25 first is the winner.

21	22 Plastic bag	23	24	25 WINNER
20	19	18 Plastic toy	17	16
11 Plastic bottle	12	13	14 Fishing hook	15
10	9 Drink can	8	7	6 Fishing net
1 START	2	3	4 Plastic lid	5

Q2e. What are some good points and weak points about Margaret's board game? How can you improve the overall visual appeal of Margaret's board game?

Good points	Weak points

Improvements

Now it's your turn.

How will you plan your game?

3. Plan...

Q3a. Use the grid to *draw a bird's-eye view of your initial board-game idea.*

Some things to consider are:

- the shape of the board game
- the size of the board game
- the number of squares
- the colours of the squares
- the types of rubbish.

Q3b. Ask your classmates to look at your board-game idea. How could it be improved? Write their comments here.

Q3c. *Consider the feedback, then draw your final idea:* add colour, label the rubbish squares and number all the squares. Add any other information and details.

Q3d. *List the materials and the quantities you will need to make the board game.*

Materials for the board game	Quantity

Procedure

Q3e. *Write down the steps you will follow to make your board game.* You can have as many steps as you like.

Q3f. What rules will you need to play your game? *Write them down.*

Tools and Equipment

Safety is important to prevent accidents and injuries. There is a safety risk when using paper, cardboard, tools and equipment. "**PPE**" (Personal Protective Equipment) is the general term used to describe safety gear. You should use PPE to protect you when using tools and equipment.

Q3g. *Write down the tools and equipment you will need to make your board game.*

Some examples of PPE: rubber gloves, face masks and safety glasses.

4. Create... [Optional]

Making a board game

You need to:

- gather all the materials, tools and equipment
- follow your procedure to make the board game
- get feedback from your teacher.

5. Improve...

The next step is to present your board-game idea to the class. Explain why animals living below water need a rubbish-free home.

Questions to ask are:

- What are the good points of my board-game idea?
- What are the weak points of my board-game idea?

Q5a. *Write your classmates' answers in the table.*

Good points	Weak points

Q5b. Consider your classmates' answers. What could you do to improve your board-game idea next time?

Now it's time to Reflect...

What have you learnt in this activity? *List three things.*
Hint: Look back at the key words.

1.

2.

3.

What would you like to know more about?

8

Life on Land

Caring for the life of all living things on the Earth is an important goal of the United Nations. People are clearing land to make way for farming, industries, cities and towns. As cities and towns grow, the **habitats** of animals such as birds are destroyed. Once trees are cut down, animals need **shelter**.

Key words in this activity:

- scale model
- shelter
- materials
- PPE
- habitat
- shape
- health hazard

Challenge
- plan a shelter for native birds

Groups like the RSPCA are concerned about the loss of habitats for birds in cities and towns. One way to increase bird numbers in your local area is to provide more habitats. Your school has asked all its students to create a shelter for a native bird. The size of the shelter needs to be suitable for the bird. The best ideas will be posted on your school's website to encourage the community to create bird shelters.

1. Ask...

Q1a. As cities and towns grow, the habitats of birds are destroyed. What does this mean?

A bird house

Q1b. Why do birds need shelter?

Here are a few examples of bird shelters.

Q1c. Compare the examples of these bird shelters. Which is best? Explain why. *Record your answers below.*

The best shelter is

A hanging bird shelter

This is the best bird shelter because

A natural log bird shelter

Q1d. Look at the bird shelter photographs. *What features make a good bird shelter?*

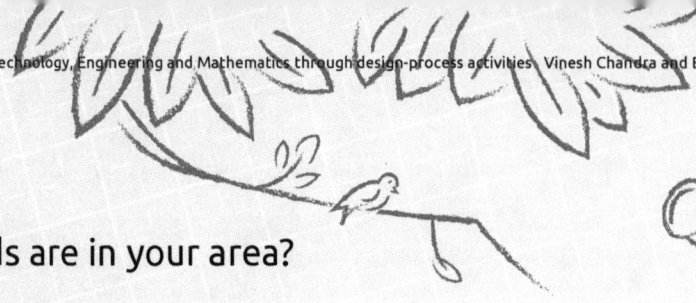

Q1e. What native birds are in your area?
List three.

1.

2.

3.

2. Imagine...

Q2a. Bird shelters can be any shape. What shapes could be used to make a bird shelter? *Sketch and label three shapes on the grids.*

Shape 1:

Shape 2:

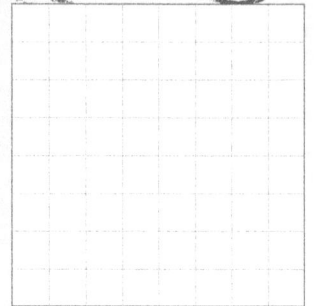

Shape 3:

Q2b. The bird shelters on page 56 are made of wood, knitted twine and a log. Which material is best suited to making a bird shelter? Why?

The best material for a bird shelter is	**This is the best material because**

To make the shelter, the material has to be *(tick all answers that apply):*

☐ scrunched

☐ twisted

☐ bent

☐ cut

☐ torn

Scrunching, twisting, bending, cutting or tearing does not change the material *(tick your answer)*.

☐ True   ☐ False

Explain your answer:

3. Plan...

Q3a. Use the grid to *draw a view of your initial bird-shelter idea.*

Q3b. Show your idea to your classmates and get feedback. Write their comments here.

Consider the feedback and review your initial idea...

Q3c. *Now draw your final idea.* Some things to consider are looks, shape, size and bird safety.

You need to:
• label the **materials**
• add any other information and details.

Challenge 8 Plan a shelter for native birds

Q3d. Complete a material list.

Part of bird shelter	Materials	Quantity

Procedure

Q3e. How could you make your bird shelter?
Draw a line from each process to the correct step number.

Step

Cut out parts.

1

Mark out parts.

2

Collect materials.

3

Paint it.

4

5

Check for quality.

6

Assemble parts.

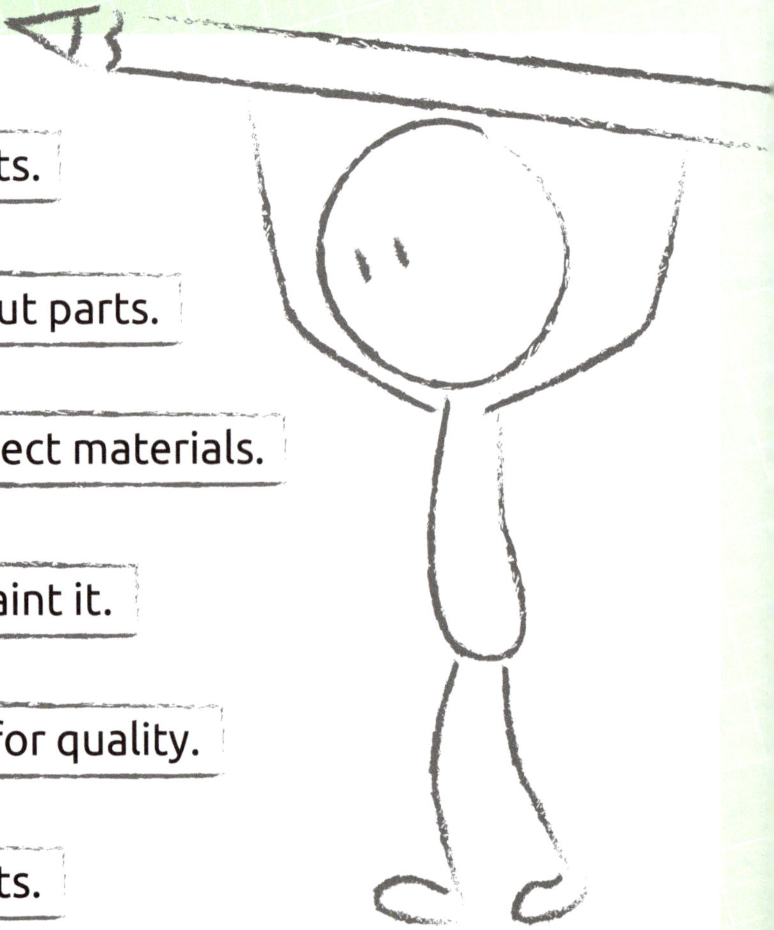

Tools and Equipment

Safety is important to prevent accidents and injuries. There is a safety risk when using tools and equipment. "**PPE**" (Personal Protective Equipment) is the general term used to describe safety gear. You should wear PPE to protect yourself when using tools and equipment.

Rubber gloves are an example of PPE.

Q3f. A number of processes can be used to make a bird shelter. These are listed in the table. *Complete the other two columns.*

Process	Health hazard	PPE I will need
Cutting straw		
Sanding wood		
Painting		
Bending		

Find-a-word   *Can you find these words in the find-a-word?*

ANIMALS
AREA
BIRD
CARE
CITY
CUT
DESTRUCTION
EARTH
ECOLOGY
FARMING
FAUNA
FEED
FLORA
GROW
HABITAT
HAZARD
HEALTH

IDEA
INDUSTRY
LAND
LIFE
LIVING
LOG
MATERIAL
NATIVE
NATURE
RSPCA
SHAPE
SHELTER
STRAW
TREE
USE
WILDLIFE
WOOD

```
U N A T U R E B F L O R A F
S T R A W O O D I L I F E A
E H E A L T H K V R N E X U
C W A H A Z A R D Q D R J N
L I V I N G B F M C U T E A
N L S W D K I A P R S P C A
A D H D E S T R U C T I O N
T L A T A E A M Z I R D L I
I I P L R Y T I C T Y E O M
V F E I T E F N A Y L A G A
E E A M H X E G R O W O Y L
J L C S H E L T E R P X G S
```

60

4. Create... [Optional]

Making a **scale model** of a bird shelter

You need to:
- gather the materials: paddle-pop/ice-cream sticks, paper and cardboard
- gather the tools: scissors, stapler and paper clips or bulldog clips
- gather the adhesives: glue and adhesive tape
- follow your procedure to make the bird shelter
- get feedback from your teacher.

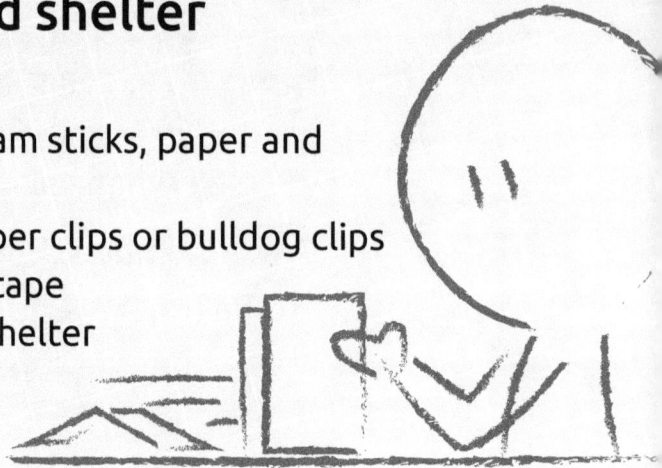

5. Improve...

The next step is to present your idea of the bird shelter to your class. Explain the process of building the shelter.

Questions to ask are:
- What are the good points of my bird-shelter idea?
- What are the weak points of my bird-shelter idea?

Q5a. *Write your classmates' answers in the table.*

Good points	Weak points

Q5b. Consider your classmates' comments. How could you improve your bird shelter next time?

Now it's time to Reflect...

What have you learnt in this activity? *List three things.*
Hint: Look back at the key words.

1.

2.

3.

What would you like to know more about?